Vanessa Grabelt

Aus der Reihe: e-fellows.net stipendiaten-wissen

e-fellows.net (Hrsg.)

Band 139

Renormalization of the regularized relativistic electron-positron field

GRIN Verlag

Bibliografische Information der Deutschen Nationalbibliothek:

Die Deutsche Bibliothek verzeichnet diese Publikation in der Deutschen Nationalbibliografie; detaillierte bibliografische Daten sind im Internet über http://dnb.d-nb.de/ abrufbar.

Imprint:

Druck und Bindung: Books on Demand GmbH, Norderstedt Germany
ISBN: 978-3-640-96724-7

This book at GRIN:

http://www.grin.com/en/e-book/175581/renormalization-of-the-regularized-relativistic-electron-positron-field

Technische Universität München

Department of Mathematics

Bachelor's Thesis

Renormalization of the regularized relativistic electron-positron field

Zusammenfassung

Diese Bachelorarbeit befasst sich mit einer Fragestellung aus der mathematischen Quantenelektrodynamik, einem wachsenden Forschungsfeld, welches auf eine saubere mathematische Behandlung der physikalischen Theorie der Quantenelektrodynamik (QED) abzielt. QED ist die relativistische Feldtheorie der Elektrodynamik, die die Quantenmechanik und die Spezielle Relativitätstheorie konsistent vereint. Mathematisch betrachtet ist die QED eine abelsche Eichtheorie mit der Symmetriegruppe U(1) (Phasenfaktoren). Das Eichfeld, welches die Wechselwirkung zwischen geladenen Spin-1/2 Feldern vermittelt ist das elektromagnetische Feld.

Inhalt dieser Abschlussarbeit ist die Renormierung des regularisierten relativistischen Elektron-Positron-Feldes zusammen mit einer Coulomb-Wechselwirkung, die mit Hilfe der Normalordnung realisiert wird. Der renormierte Hamilton-Operator unterscheidet sich von dem ursprünglichen durch einen quadratischen Term. Durch die Wahl einer geeigneten Normalordnung, welche einer nicht-perturbativen Neudefinition der Einteilchen-Zustände entspricht, kann der zusätzliche auftretende Term im Zuge des Renormierungs- prozesses absorbiert werden. Am Ende der Arbeit werden verschiedene mögliche Interpretationen und Implikationen der Renormierung betrachtet.

Contents

1 Motivation

This thesis is motivated by questions arising in the field of Mathematical Quantum Electrodynamics, the attempt of a proper mathematical description of Quantum Electrodynamics (QED). QED is the relativistic quantum field theory of electrodynamics, which unifies Quantum Mechanics and Special Relativity in a consistent manner. From the mathematical point of view, QED is an abelian gauge theory with the symmetry group U(1) (phase factors) and the gauge field mediating the interaction between the charged spin-1/2 fields is the electromagnetic field.

Aim of this thesis is the renormalization of the regularized relativistic electron positron field together with a Coulomb interaction. The idea of the renormalization procedure is to compare the normal-ordered Hamiltonian with the original one. Choosing a conventional normal ordering, the change in Hamiltonian is given by a quadratic term. The choice of a suitable normal ordering amounts in a non-perturbative redefinition of the electron/positron states. This allows for the interpretation of change in the Hamiltonian as a certain renormalization. The proper interpretation of this renormalization and the various implications are discussed at the end.

2 Basics of QED

2.1 Why Fields?

QED is the relativistic quantum field theory of electrodynamics, which unifies Quantum Mechanics and Special Relativity in a consistent manner. The two most important principles in QED which, according to Weinberg [5] essentially make the introduction of fields natural, are Poincare Covariance and the Cluster Decomposition Principle. Poincare Covariance arises from Special Relativity, whereas the Cluster Decomposition Principle guarantees some 'locality', meaning that two measurements sufficiently separated should not depend on each other.

2.2 Fields

In QED fields are the basic objects one has to deal with. A standard way to obtain the commonly used fields in QED, namely scalar fields and Dirac (spinor) fields, is the canonical quantization of classical fields in Minkowski space. Another method is the path integral quantization, which is of no particular importance for our case and therefore won't be discussed here.

2.3 Canonical Quantization

In the process of canonical quantization one starts with a classical field ϕ and its Lagrangian density, which is a function of the field and its derivatives. The Lagrangian density is connected with the Lagrange function, which is a functional of the field via

$$L[\phi] = \int dx \; \mathscr{L}\Big(\phi(x,t), \dot{\phi}(x,t), \partial_x\phi(x,t)\Big) \tag{1}$$

Using the Lagrangian density one directly obtains the conjugated momentum field π, which is in direct correspondence to the conjugated momentum in classical mechanics.

$$\pi(x,t) = \frac{\partial \mathscr{L}}{\partial \dot{\phi}}(x,t) \tag{2}$$

Now one promotes the classical fields ϕ, π to operators Φ, Π and requires the canonical (anti)commutation relations well-known from quantum mechanics. If commutation (bosonic) oder anticommutation (fermionic) rules should be used, can be determined using the spin statistics theorem. The type of the field determines the particle type (scalar fields $\rightarrow$ bosons, Dirac fields $\rightarrow$ fermions).

$$\begin{aligned}
[\Phi(x,t), \Phi(y,t)]_{\pm} &= 0 \\
[\Pi(x,t), \Pi(y,t)]_{\pm} &= 0 \\
[\Phi(x,t), \Pi(y,t)]_{\pm} &= i\hbar\delta(x-y)
\end{aligned}$$

Exploiting the fact that the Hamiltonian density is connected with the Lagrangian density via Legendre transformation, the Hamilton operator can be easily determined.

$$H(t) = \int dx \; \Big(\Pi(x,t)\dot{\Phi}(x,t) - \mathscr{L}\Big) \tag{3}$$

2.4 Renormalization

In quantum field theories one often encounters divergent integrals in calculations. In general it can be said that when describing spacetime as a continuum, some constructions (e.g. from quantum mechanics) are ill-defined.
A renormalization program was first developed in quantum electrodynamics (QED), when facing infinite integrals in perturbation theory calculations. Infinities in quantum field theories often arise due to so-called virtual particle loop-integration (the term loop comes from the Feynman diagrammatic description). Virtual particles, in contrast to real particles, are said to be off-(mass)shell meaning that they don't fulfill the relativistic energy-momentum relation with a certain mass m: $m^2c^4 = E^2 - p^2c^2$. The corresponding loops in Feynman diagrams describe closed lines, which have no constraint on possible momenta q, therefore one has to integrate over all momenta q.
From the physical point of view every quantum field theory (with interactions) is an effective theory and valid only when considering a finite scope of energy. Therefore perturbative calculations including integrations over momenta out of this scope yield meaningless infinite results.

2.4.1 Bare and renormalized parameters

Renormalization determines the relationship between parameters in the theory, when the parameters describe large distance scales (probed with particles of small momenta) differ from the parameters describing small distances (probed with particles of high momenta). The theoretical so called bare parameters appearing (e.g. bare mass m_0, fine structure constant α) don't correspond to the parameters measured in the laboratory (e.g. physical mass m_{ph}, fine structure constant α_{ph}).
In order to obtain physically relevant constants the virtual-particle loop effects have to be taken into account and have to be 'absorbed' into the parameters. This seems to be a problem since divergent integrals would require divergent bare parameters in order to recover finite physical parameters. To circumvent this problem first a regularization procedure is performed, which separates contributions from inside and outside the corresponding energy scope.

Regularization In order to deal with infinities in a proper mathematical manner and adress the principle of effective theories, the loop integrals are regularized by inserting a regulator in the integrand. This regulator has a characteristic energy scale (cutoff, in momentum space Λ) and drops off fast enough at energies high above this value. This ensures convergence of the integrals to finite cutoff-dependent terms and recovers the original divergence if the cutoff goes to infinity.
At this point we have to emphasize that the bare parameters secretly depend on the characteristic energy scale, since we always describe physical quantities which don't depend on our cutoff explicitly (it shouldn't matter if we choose 2Λ or $1/2\Lambda$ instead of Λ). Now one can cancel out some correction terms with contributions from cutoff-dependent counterterms and afterwards take the cutoff back to infinity such that finite physical results are recovered.

Renormalizability It turns out that corrections arising from energies above the cutoff can be described by redefined parameters, like the mass or coupling constant. These corrections grow as the concerned energy differs from the introduced energy scope. If a finite number of redefined parameters suffices (equivalent: if there is only a finite number of irreducible diagrams), then the theory is said to be renormalizable.
QED is a renormalizable theory with three parameters which have to be redefined, namely charge of the electron e (or $\alpha = e^2$), mass of the electron m and field normalization Z.

2.4.2 First-order radiative corrections in QED

In QED there are three basic examples for divergent loop integrals which are the first order radiative corrections, namely vacuum polarization, self-energy and vertex correction. These effects and the different renormalizations adressing them will be discussed briefly.

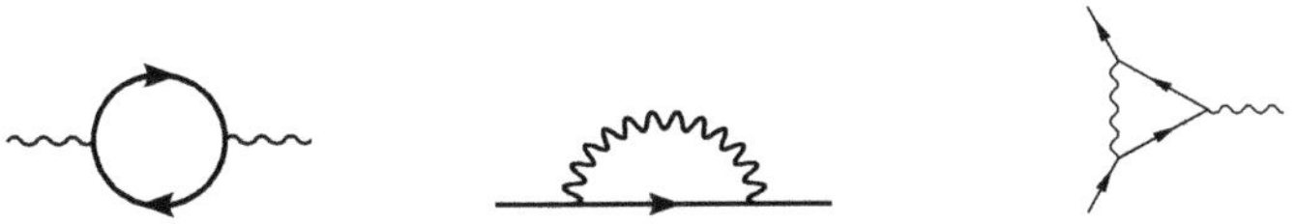

Figure 1: a) vacuum polarization, b) self-energy, c) vertex correction

Vacuum polarization and charge renormalization In a background electric field (for example caused by an electron) short-living virtual electron-positron pairs may be produced. The Feynman diagram represents a photon creating a virtual electron-positron pair which then annihilates to a photon. These pairs can be seen to act like dipols and therefore partially screen the field (for an electron the charge) due to their rearrangement. The renormalization of this divergent integral is therefore often called charge renormalization. The physical quantity associated to this renormalization is the fine structure constant α_{ph}.

Self-energy and mass renormalization Self-energy describes the amount of energy an electron has due to interaction with the field it is part of, it is often adressed as self-interaction. The Feynman diagram depicts an electron which quickly emits and reabsorbs a virtual photon, where the photon is the carrier of the electromagnetic force. The renormalization of this divergent integral is often called mass renormalization since the self-interaction amounts to an increased mass. The physical quantity associated to this renormalization is the mass m_{ph}.

Vertex-correction and field (wavefunction) renormalization The one-loop correction to the vertex function describes the coupling between the external field caused by a vectorpotential (photon) and an electron beyond leading order of perturbation theory. This is the dominant contribution to the anomalous magnetic moment of the electron. The physical quantity associated to this renormalization is the field renormalization Z (often called wave function renormalization). In the Feynman diagram an electron emits a virtual photon, emits a second real photon, and reabsorbs the first one.

2.4.3 Normal ordering

Normal order Being in normal order amounts to saying that all creation operators are to the left of all annihilation operators. No further specification is needed since all annihilation operators commute with one another as do all annihilation operators.
In particular vacuum expectation values of any normal ordered product of creation and annihilation operators are zero, since an annihilation operator applied to the vacuum state $|0\rangle$ gives zero. Of particular interest are therefore normal-ordered Hamiltonians with zero ground state energy ($\langle 0|\hat{H}|0\rangle = 0$).

Normal ordering Normal ordering (or Wick ordering) is the process to (anti)commute all creation operators to the left while ignoring non-vanishing (anti)commutators. Ignoring non-vanishing (anti)commutators corresponds to throwing away all diagrams with self-loops. Normal ordering of an operator A is denoted by $:A:$.

Normal ordering as renormalization Normal ordering is the simplest form of a renormalization. It is of particular importance when quantizing a classical Hamiltonian. On the one hand classically everything commutes, for that reason the operator order in the first place seems to have a freedom of choice. On the other hand different orderings of operators lead to differences in the ground state energy. Therefore by normal ordering the energy is shifted by an infinite constant (ground state energy), which is physically motivated by the fact that additive constants only lead to phase factors in the time evolution being not measurable.
Since normal ordering amounts to formally 'throwing away' all diagrams with self-loops which produce divergencies, by taking the difference of the original operator and the normal ordered one a renormalization procedure is defined.

3 Mathematical description

The definitions underlying this thesis are following basically [2], except the fact that here no external fields are included.

Hilbert space For the electron-positron field the corresponding Hilbert space is the space of $L^2(\mathbb{R}^3)$ spinors, namely $\mathfrak{h} = L^2(\mathbb{R}^3) \otimes \mathbb{C}^4$.

One-particle spaces The one-electron space is defined by a closed subspace $\mathfrak{F}_+^1 := \mathfrak{h}_+$ of $\mathfrak{h}$. The Hilbert space $\mathfrak{h} = \mathfrak{h}_+ \oplus \mathfrak{h}_-$ is an orthogonal sum of $\mathfrak{h}_+$ and its orthogonal complement $(\mathfrak{h}_+)^\perp = \mathfrak{h}_-$. The one-positron subspace $\mathfrak{F}_-^1 := C\mathfrak{h}_-$ is given by the charge conjugate C of $\mathfrak{h}_-$. Charge conjugation operates in position space as $(C\Psi)(x) = i\beta\alpha_2\overline{\Psi(x)}$ and in momentum space as $(C\hat{\Psi})(p) = i\beta\alpha_2\overline{\hat{\Psi}(-p)}$, where α_2 and β are 4x4 matrices contained in the free Dirac operator and will be defined later on. $P_\pm$ denote the orthogonal projectors onto the subspaces $\mathfrak{h}_\pm$.

Fock space The n-particle electron subspace is $\mathfrak{F}_+^n = \bigwedge_{\nu=1}^n \mathfrak{h}_+$ and the m-particle positron subspace is $\mathfrak{F}_-^m = \bigwedge_{\nu=1}^m C\mathfrak{h}_-$. The zero particle subspaces (vacua) are the same $\mathfrak{F}_+^o = \mathbb{C} = \mathfrak{F}_-^o$. The total Fock space is given by a direct sum of product spaces containing n electrons and m positrons

$$\mathfrak{F} = \bigoplus_{n,m=0}^{\infty} \underbrace{(\mathfrak{F}_+^n \otimes \mathfrak{F}_-^m)}_{:=\mathfrak{F}^{(n,m)}}$$

Annihilation & Creation operators On the Fock space one can define electron and positron annihilation and creation operators. Let $x = (\vec{x}, \sigma) \in \mathbb{R}^3 \times \{1,2,3,4\} := G$ denote the space-spin point and dx the product measure of Lebesgue measure in 1st and counting measure in 2nd factor, namely $\int \mathrm{dx} = \sum_\sigma \int \mathrm{d}^3\mathrm{x}$. Let f and g be functions in the Hilbert space $\mathfrak{h}$.
The electron annihilation operator $a(f) : \mathfrak{F}^{(n+1,m)} \to \mathfrak{F}^{(n,m)}$ annihilates an electron in the state P_+f, according terms hold for the other operators $(a^*(f), b(g), b^*(g))$ as well

$$(a(f)\Psi)(x_1, ..., x_n; y_1, ..., y_m) = \sqrt{n+1} \int_G \mathrm{dx}\, \overline{(P_+f)(x)}\Psi(x, x_1, ..., x_n; y_1, ..., y_m)$$

The electron creation operator $a^*(f) : \mathfrak{F}^{(n-1,m)} \to \mathfrak{F}^{(n,m)}$ creates an electron

$$(a^*(f)\Psi)(x_1, ..., x_n; y_1, ..., y_m) = \frac{1}{\sqrt{n}} \sum_{j=1}^{n} (-1)^{j+1} (P_+f)(x_j)\Psi(x_1, ..., \hat{x}_j, ..., x_n; y_1, ..., y_m)$$

The operator $a^*(f)$ is the adjoint of $a(f)$ with the following anticommutation relations

$$\{a(f), a(g)\} = 0 = \{a^*(f), a^*(g)\}$$
$$\{a(f), a^*(g)\} = (f, P_+g)$$

The positron annihilation operator $b(g) : \mathfrak{F}^{(n,m+1)} \to \mathfrak{F}^{(n,m)}$ annihilates a positron

$$(b(g)\Psi)(x_1, ..., x_n; y_1, ..., y_m) = (-1)^n \sqrt{m+1} \int_G \mathrm{dy}\, \overline{(CP_{-}g)(y)} \Psi(x_1, ..., x_n; y, y_1, ..., y_m)$$

The positron creation operator $b^*(g) : \mathfrak{F}^{(n,m-1)} \to \mathfrak{F}^{(n,m)}$ creates a positron

$$(b^*(g)\Psi)(x_1, ..., x_n; y_1, ..., y_m) = \frac{(-1)^n}{\sqrt{m}} \sum_{k=1}^{m} (-1)^{k+1} (CP_{-}g)(y_k) \Psi(x_1, ..., x_n; y_1, , ..., \hat{y}_k ..., y_m)$$

The operator $b^*(g)$ is the adjoint of $b(g)$ with the following anticommutation relations

$$\{b(f), b(g)\} = 0 = \{b^*(f), b^*(g)\}$$
$$\{b^*(f), b(g)\} = (f, P_{-}g)$$

By linearity all operators extend to bounded operators on $\mathfrak{F}$, note that $f \to a(f)$ and $g \to b^*(g)$ are antilinear and $f \to a^*(f)$ and $g \to b(g)$ are linear.
The annihilation operators a(f) and b(f) both annihilate the vacuum vector (for arbitrary f) $|0\rangle = 1$, which is unique up to a phase

$$a(f)|0\rangle = 0 = b(f)|0\rangle$$

In addition to the anticommutation relations between a and a^* resp. b and b^*, all mixed anticommutators vanish.

$$\{a(f), b(g)\} = 0 = \{a^*(f), b^*(g)\}$$
$$\{a(f), b^*(g)\} = 0 = \{a^*(f), b(g)\}$$

Using orthonormal bases for $\mathfrak{h}_\pm$ operators creating/annihilating electron/positron at a space-spin point x can be defined. Let f_ν be an orthonormal basis for $\mathfrak{h}_+$ with $\nu \geq 0$ and for $\mathfrak{h}_-$ with $\nu < 0$, then define

$$a(x) = \sum_{\nu=0}^{\infty} a(f_\nu) f_\nu(x) \qquad b(x) = \sum_{\nu=-\infty}^{-1} b(f_\nu) \overline{f_\nu(x)}$$

The commutation relations from above then give, with $P_\pm(x, y)$ as integral kernels of $P_\pm$

$$\{a(x), a(y)\} = \{a^*(x), a^*(y)\} = 0 = \{b(x), b(y)\} = \{b^*(x), b^*(y)\}$$
$$\{a(x), b(y)\} = \{a^*(x), b^*(y)\} = 0 = \{a(x), b^*(y)\} = \{a^*(x), b(y)\}$$
$$\{a(x), a^*(y)\} = P_+(x, y) \qquad \{b^*(x), b(y)\} = P_-(x, y)$$

Dirac Operator The free Dirac Operator acting on the Hilbert space $\mathfrak{h}$ is given by

$$D_0 = \vec{\alpha} \cdot \vec{p} + m_0 \beta \quad \vec{p} = \frac{1}{i} \vec{\nabla}$$

Where α and β are the 4×4 Dirac matrices, which are in standard representation given by the 2×2 Pauli matrices $\sigma_{1,2,3}$ and the identity $\mathbb{1}_2$

$$\vec{\alpha} = \begin{pmatrix} 0 & \vec{\sigma} \\ \vec{\sigma} & 0 \end{pmatrix} \quad \beta = \begin{pmatrix} \mathbb{1}_2 & 0 \\ 0 & -\mathbb{1}_2 \end{pmatrix}$$

$$\sigma_1 = \begin{pmatrix} 0 & 1 \\ 1 & 0 \end{pmatrix} \quad \sigma_2 = \begin{pmatrix} 0 & -i \\ i & 0 \end{pmatrix} \quad \sigma_3 = \begin{pmatrix} 1 & 0 \\ 0 & -1 \end{pmatrix}$$

The number m_0 appearing in the Dirac operator adresses the bare mass of the electron/positron.

Field Operators The field operator Ψ describes a Dirac field which cannot be interpreted as one particle field which is due to the problem of negative energies in the Dirac equation.

$$i\partial_t \Psi(\vec{x}, t) = (\vec{\alpha} \cdot \vec{p} + m_0\beta)\Psi(\vec{x}, t)$$

Due to this the concept of antiparticles was introduced, here the positron is the antiparticle of the electron. This is the main reason why 2 types of creation and annihilation operators are needed. Therefore Ψ contains an electron annihilation and a positron creation operator, for its adjoint Ψ^* vice versa.

$$\Psi(x) = a(x) + b^*(x) \qquad \Psi^*(x) = a^*(x) + b(x)$$

Since the solutions of the Dirac equation describe spinors, it is more precise to call Ψ an operator valued spinor. Both field operators Ψ and Ψ^* are bounded operators on the Fock Space $\mathfrak{F}$.

Free Hamiltonian The Hamiltonian of the free electron-positron field is connected with the free Dirac operator. Recall that : : denotes normal ordering.

$$\mathbb{H}_0 = \int d^3x \ :\Psi^*(x) D_0 \Psi(x): \tag{4}$$

Coulomb interaction The electron positron field introduced above can interact with itself due to Coulomb interaction. The Coulomb potential is a symmetric interaction potential $W(x, y) = \delta_{\sigma,\tau} \frac{1}{|\vec{x}-\vec{y}|}$. In best analogue to the classical energy of a field the interaction is only partially normal ordered. This is due to the physical motivation that the density-density interaction should be positive, which is no longer true after renormalization.

$$\mathbb{W}^{bare} = \int d^3x \int d^3y \ W(x, y) :\Psi^*(x)\Psi(x)::\Psi^*(y)\Psi(y): \tag{5}$$

Regularization Since the potential W(x,y) is clearly singular and would cause positive singularities in $\mathbb{W}^{bare}$ we will consider regularized versions. Therefore the field operators will be cut off above momenta of $\Lambda > 0$, meaning that

$$\Psi(x) = \frac{1}{(2\pi)^{3/2}} \int_{|\vec{p}|<\Lambda} d^3p \ \hat{\Psi}(\vec{p}, \sigma) e^{-i\vec{p}\cdot\vec{x}}$$

4 Calculations

The guideline for this chapter is as follows:

Bare and normal ordered Hamiltonian As a first step the difference between the bare (unrenormalized) Hamiltonian and the renormalized one is determined. In the next step a non-perturbative choice of the one particle states is made in order to organize the arising terms such that they can be described by an operator similar to a Dirac operator.

Dressed electron The following part is devoted to determining the operator of Dirac kind and conditions under which this identification is possible. This determines the new non-perturbative one electron states, which is why this section is called dressed electron. Using a fixed point argument, the solutions will be determined for the two cases, of zero and nonzero bare mass m_0.

Properties of renormalized Hamiltonian In the last part the properties of the new Hamiltonian will be investigated. This is necessary in order to decide whether and in which fashion the Coulomb interaction can be included in the renormalization program.

4.1 Fully normal ordered interaction

4.1.1 Bare and normal ordered Hamiltonian

First we recall the most important facts from section 3 concerning the field operators and the non vanishing anticommutators between the creation & annihilation operators, which have to be used for normal ordering.

$$\Psi(x) := a(x) + b^*(x) \qquad \Psi^*(x) := a^*(x) + b(x)$$

$$\{a(x), a^*(y)\} = P_+(x,y) \qquad \{b^*(x), b(y)\} = P_-(x,y) \qquad \text{others} = 0$$

The bare interaction energy is only partially normal ordered, its renormalization will be determined here by carrying out the full normal ordering. This entails determining the splitting of $\mathfrak{h}$ which is denoted by subscript α.

$$\mathbb{W}^{bare} = \frac{\alpha}{2}\int d^3x \int d^3y\, W(x,y)\, \overbrace{:\Psi^*(x)\Psi(x)::\Psi^*(y)\Psi(y):}^{w^{bare}}$$
$$\mathbb{W}^{ren}_\alpha = \frac{\alpha}{2}\int d^3x \int d^3y\, W(x,y)\, \underbrace{:\Psi^*(x)\Psi^*(y)\Psi(y)\Psi(x):}_{w^{ren}_\alpha}$$

$$\begin{aligned} :\Psi^*(x)\Psi(x)::\Psi^*(y)\Psi(y): &:=: [a^*(x)+b(x)][a(x)+b^*(x)] :: [a^*(y)+b(y)][a(y)+b^*(y)] : \\ &=: a^*(x)a(x) + a^*(x)b^*(x) + b(x)a(x) + b(x)b^*(x) : \\ &\quad : a^*(y)a(y) + a^*(y)b^*(y) + b(y)a(y) + b(y)b^*(y) : \end{aligned}$$

$$
\begin{aligned}
:\Psi^*(x)\Psi^*(y)\Psi(y)\Psi(x): &= :[a^*(x)+b(x)][a^*(y)+b(y)][a(y)+b^*(y)][a(x)+b^*(x)]: \\
&= :\big(a^*(x)a^*(y)+a^*(x)b(y)+b(x)a^*(y)+b(x)b(y)\big) \\
&\quad \big(a(y)a(x)+a(y)b^*(x)+b^*(y)a(x)+b^*(y)b^*(x)\big):
\end{aligned}
$$

In the following the results of the corresponding normal orderings are tabulated. The numbers on the left and right indicate which expressions are related and will cancel each other using the anticommutation relations.

	w^{bare}					w_α^{ren}				
1	+	$a^*(x)$	$a(x)$	$a^*(y)$	$a(y)$	+	$a^*(x)$	$a^*(y)$	$a(y)$	$a(x)$	1
2	+	$a^*(x)$	$a(x)$	$a^*(y)$	$b^*(y)$	-	$a^*(x)$	$a^*(y)$	$b^*(x)$	$a(y)$	5
3	+	$a^*(x)$	$a(x)$	$b(y)$	$a(y)$	+	$a^*(x)$	$a^*(y)$	$b^*(y)$	$a(x)$	2
4	-	$a^*(x)$	$a(x)$	$b^*(y)$	$b(y)$	+	$a^*(x)$	$a^*(y)$	$b^*(y)$	$b^*(x)$	6
5	+	$a^*(x)$	$b^*(x)$	$a^*(y)$	$a(y)$	+	$a^*(x)$	$b(y)$	$a(y)$	$a(x)$	3
6	+	$a^*(x)$	$b^*(x)$	$a^*(y)$	$b^*(y)$	+	$a^*(x)$	$b^*(x)$	$b(y)$	$a(y)$	7
7	+	$a^*(x)$	$b^*(x)$	$b(y)$	$a(y)$	-	$a^*(x)$	$b^*(y)$	$b(y)$	$a(x)$	4
8	-	$a^*(x)$	$b^*(x)$	$b^*(y)$	$b(y)$	+	$a^*(x)$	$b^*(y)$	$b^*(x)$	$b(y)$	8
9	+	$b(x)$	$a(x)$	$a^*(y)$	$a(y)$	-	$a^*(y)$	$b(x)$	$a(y)$	$a(x)$	9
10	+	$b(x)$	$a(x)$	$a^*(y)$	$b^*(y)$	-	$a^*(y)$	$b^*(x)$	$b(x)$	$a(y)$	13
11	+	$b(x)$	$a(x)$	$b(y)$	$a(y)$	+	$a^*(y)$	$b^*(y)$	$b(x)$	$a(x)$	10
12	-	$b(x)$	$a(x)$	$b^*(y)$	$b(y)$	-	$a^*(y)$	$b^*(y)$	$b^*(x)$	$b(x)$	14
13	-	$b^*(x)$	$b(x)$	$a^*(y)$	$a(y)$	+	$b(x)$	$b(y)$	$a(y)$	$a(x)$	11
14	-	$b^*(x)$	$b(x)$	$a^*(y)$	$b^*(y)$	-	$b^*(x)$	$b(x)$	$b(y)$	$a(y)$	15
15	-	$b^*(x)$	$b(x)$	$b(y)$	$a(y)$	+	$b^*(y)$	$b(x)$	$b(y)$	$a(x)$	12
16	+	$b^*(x)$	$b(x)$	$b^*(y)$	$b(y)$	+	$b^*(y)$	$b^*(x)$	$b(x)$	$b(y)$	16

Subtracting the corresponding terms gives different combinations of projection kernels and creation & annihilation operators.

	$w^{bare} - w_\alpha^{ren}$
1	$P_+(x,y)a^*(x)a(y)$
2	$P_+(x,y)a^*(x)b^*(y)$
3-8	0
9	$P_+(x,y)b(x)a(y)$
10	$-P_+(x,y)b^*(y)b(x) - P_-(y,x)a^*(y)a(x) + P_-(y,x)P_+(x,y)$
11	0
12	$P_-(y,x)a(x)b(y)$
13	0
14	$P_-(y,x)b^*(x)a^*(y)$
15	0
16	$P_-(y,x)b^*(x)b(y)$

$$
\begin{aligned}
w^{bare} - w_\alpha^{ren} &= P_+(x,y)\left[a^*(x)a(y)+a^*(x)b^*(y)-a(y)b(x)-b^*(y)b(x)\right] \\
&\quad - P_-(y,x)\left[a^*(y)a(x)+a^*(y)b^*(x)-a(x)b(y)-b^*(x)b(y)\right] \\
&\quad + P_+(x,y)P_-(y,x) \\
&= P_+(x,y) :\Psi^*(x)\Psi(y): - P_-(y,x) :\Psi^*(y)\Psi(x): + P_+(x,y)P_-(y,x)
\end{aligned}
$$

4.1.2 Interim result for connection between bare and normal ordered Hamiltonian

$$\begin{aligned}
\mathbb{W}^{bare} - \mathbb{W}^{ren}_{\alpha} &= \frac{\alpha}{2}\int d^3x \int d^3y \left(\frac{P_+(x,y)}{|\vec{x}-\vec{y}|} : \Psi^*(x)\Psi(y) : - \frac{P_-(y,x)}{|\vec{x}-\vec{y}|} : \Psi^*(y)\Psi(x) :\right) \\
&+ \frac{\alpha}{2}\int d^3x \int d^3y \; \frac{P_+(x,y)P_-(y,x)}{|\vec{x}-\vec{y}|} \\
&= \frac{\alpha}{2}\int d^3x \int d^3y \; \left(\frac{P_+(x,y) - P_-(x,y)}{|\vec{x}-\vec{y}|} : \Psi^*(x)\Psi(y) : + \frac{P_+(x,y)P_-(y,x)}{|\vec{x}-\vec{y}|}\right) \\
\mathbb{H}^{bare} &= \mathbb{H}_0 + \mathbb{W}^{bare} = \mathbb{H}_0 + (\mathbb{W}^{bare} - \mathbb{W}^{ren}_{\alpha}) + \mathbb{W}^{ren}_{\alpha} \\
&= \int d^3x \; : \Psi^*(x) D_0 \Psi(x) : + \frac{\alpha}{2}\int d^3x \int d^3y \; \frac{P_+(x,y) - P_-(x,y)}{|\vec{x}-\vec{y}|} : \Psi^*(x)\Psi(y) : \\
&+ \mathbb{W}^{ren}_{\alpha} + \frac{\alpha}{2}\int d^3x \int d^3y \; \frac{P_+(x,y)P_-(y,x)}{|\vec{x}-\vec{y}|}
\end{aligned}$$

It is desirable to determine a normal ordered Hamiltonian containing a modified Dirac operator $D_{Z,m}$ with a mass renormalization (physical mass) $m > m_0$ and a wavefunction renormalization $Z < 1$.

$$D_{Z,m} = Z^{-1}(\vec{\alpha}\cdot\vec{p} + m\beta) \tag{6}$$

This operator has to be obtained by introducing non-perturbative one-electron states.

$$\mathbb{H}^{ren}_{\alpha} = \int d^3x \; : \Psi^*(x) D_{Z,m} \Psi(x) : + \mathbb{W}^{ren}_{\alpha} \tag{7}$$

In the following we want to show that by a suitable choice of Z, m and the electron space $\mathfrak{h}_+$ there is asymptotic equality between the original Hamiltonian $\mathbb{H}^{bare}$ and the normal ordered Hamiltonian $\mathbb{H}^{ren}_{\alpha}$ for small momenta ($|\vec{p}| \ll \Lambda$).

$$\mathbb{H}^{bare} = \mathbb{H}^{ren}_{\alpha} \tag{8}$$

4.1.3 Cutoff dependent constant

In this section we turn our attention to the term which doesn't involve the field operators Ψ and Ψ^* and therefore is constant. This constant can be dropped since it has no physically relevant effects.

$$\frac{\alpha}{2}\int d^3x \int d^3y \; \frac{P_+(x,y)P_-(y,x)}{|\vec{x}-\vec{y}|}$$

We will show that this constant is positive and (cutoff-dependent) infinite.

Positivity In order to show positivity we will use a formula found by Feffermann-de la Llave (see [8]) and the characteristic property for projectors which translates to a corresponding property for their integral kernels. $P_\pm(\cdot,\cdot)$ are integral kernels of projectors $P_\pm$ with the basic property $P_\pm^2 = P_\pm$. Therefore they fulfill

$$(P_\pm^2 f)(x) = (P_\pm f)(x) \quad \Rightarrow \quad P_\pm(x,y) = \int d^3z \; P_\pm(x,z)P_\pm(z,y) \tag{9}$$

The Feffermann-de la Llave formula allows for writing the potential energy in a Coulomb field as an integral over spheres ($B(z,R)$ denotes a sphere with radius R around z).

$$\frac{1}{|x-y|} = \frac{1}{\pi}\int d^3z \int dR \,\frac{\chi_{x,y\in B(z,R)}}{R^5} = \frac{1}{\pi}\int d^3z \int dR \,\frac{\chi_{x\in B(z,R)}\chi_{y\in B(z,R)}}{R^5} \tag{10}$$

Using this two properties (9) and (10) as well as Fubini's theorem one can show, that the integral is strictly positive.

$$\begin{aligned}
&\frac{\alpha}{2}\int d^3x \int d^3y \,\frac{P_+(x,y)P_-(y,x)}{|\vec{x}-\vec{y}|}\\
=&\frac{1}{\pi}\int d^3x \int d^3y \int d^3z \int dR \,\frac{P_+(x,y)P_-(y,x)\chi_{x\in B(z,R)}\chi_{y\in B(z,R)}}{R^5}\\
=&\frac{1}{\pi}\int d^3x \int d^3y \int d^3z \int dR \int d^3u \,\frac{P_+(x,y)P_-(y,u)P_-(u,x)\chi_{x\in B(z,R)}\chi_{y\in B(z,R)}}{R^5}\\
=&\frac{1}{\pi}\int d^3x \int d^3y \int d^3z \int dR \int d^3u \,\frac{P_-(u,x)\chi_{x\in B(z,R)}P_+(x,y)\chi_{y\in B(z,R)}P_-(y,u)}{R^5}\\
=&\frac{1}{\pi}\int d^3z \int dR \int d^3u \int d^3x \int d^3y \,\frac{P_-(u,x)\chi_{x\in B(z,R)}P_+(x,y)\chi_{y\in B(z,R)}P_-(y,u)}{R^5}
\end{aligned}$$

The integrand in the last line contains a partial trace of the positive operator P_+

$$P_-(x,u)^*\chi_{x\in B(z,R)}P_+(x,y)\chi_{y\in B(z,R)}P_-(y,u)$$

and is clearly positive which implies that the integrals are also all positive.

$$\frac{\alpha}{2}\int d^3x \int d^3y \,\frac{P_+(x,y)P_-(y,x)}{|\vec{x}-\vec{y}|} > 0 \tag{11}$$

Dependence on cut-off On the other hand by assuming translational invariance of $P_\pm$ it can be shown that this term is infinite and diverges proportionally to $V\Lambda^4$ where V is the volume and Λ the momentum cut-off.
For the following calculation we use that the integral kernel of the projectors $P_\pm$ in momentum space are given by $P_\pm(\vec{p}) = \frac{1}{2} + \ldots$ and therefore their fourier transform in position space is governed by a delta function $P_\pm(\vec{x}) \propto \delta(\vec{x}) + \ldots$. In order to take the momentum cutoff into account, the regularized delta function is used

$$\delta(\vec{x}) \propto \int_{|\vec{p}|<\Lambda} d^3p \; e^{i\vec{p}\cdot\vec{x}} \propto \frac{1}{|\vec{x}|}\int_0^\Lambda dp \; p\sin(p|\vec{x}|) \propto \frac{\sin(\Lambda|\vec{x}|) - \Lambda|\vec{x}|\cos(\Lambda|\vec{x}|)}{|\vec{x}|^3}$$

$$\begin{aligned}
\int d^3x \int d^3y \,\frac{P_+(x,y)P_-(y,x)}{|\vec{x}-\vec{y}|} &= \int d^3x \int d^3y \,\frac{P_+(x)P_-(-x)}{|\vec{x}|}\\
&\propto V\int d^3x \,\frac{1}{|\vec{x}|^7}\big[\sin(\Lambda|\vec{x}|) - \underbrace{\Lambda|\vec{x}|}_{:=\tilde{x}}\cos(\Lambda|\vec{x}|)\big]^2\\
&\propto V\Lambda^4 \underbrace{\int d\tilde{x}\,\frac{1}{\tilde{x}^5}\big[\sin(\tilde{x}) - \tilde{x}\cos(\tilde{x})\big]^2}_{=\text{const}}
\end{aligned}$$

The last integral behaves properly for small and large $\tilde{x}$. For small $\tilde{x}$ one obtains

$$\frac{1}{\tilde{x}^5}\Big[\underbrace{\sin(\tilde{x})}_{\approx\tilde{x}-\frac{1}{3!}\tilde{x}^3}-\tilde{x}\underbrace{\cos(\tilde{x})}_{\approx 1-\frac{1}{2!}\tilde{x}^2}\Big]^2\approx\frac{1}{\tilde{x}^5}\left[\frac{1}{3}\tilde{x}^3\right]^2\propto\tilde{x}$$

For large $\tilde{x}$ the integrand is bounded from above by

$$\frac{1}{\tilde{x}^5}\underbrace{\left[\sin(\tilde{x})-\tilde{x}\cos(\tilde{x})\right]^2}_{\approx[\tilde{x}\cos(\tilde{x})]^2}\lessapprox\frac{1}{\tilde{x}^5}\left[\tilde{x}\right]^2\propto\frac{1}{\tilde{x}^3}$$

This finally proves

$$\frac{\alpha}{2}\int d^3x\int d^3y\;\frac{P_+(x,y)P_-(y,x)}{|\vec{x}-\vec{y}|}\propto V\Lambda^4 \tag{12}$$

4.1.4 Terms with one particle operators

In this subsection we turn our attention to the integral involving one particle operators which arose in the normal ordering process.

$$\frac{\alpha}{2}\int d^3x\int d^3y\;\frac{P_+(x,y)-P_-(x,y)}{|\vec{x}-\vec{y}|}:\Psi^*(x)\Psi(y):$$

We want to find an operator A with integral kernel A(x,y) such that

$$\begin{aligned}\mathbb{A}&=\int d^3x\int d^3y\;A(x,y):\Psi^*(x)\Psi(y):\\&=\int d^3x\;:\Psi^*(x)D_0\Psi(x):+\frac{\alpha}{2}\int d^3x\int d^3y\;\frac{P_+(x,y)-P_-(x,y)}{|\vec{x}-\vec{y}|}:\Psi^*(x)\Psi(y):\end{aligned}$$

We would like to obtain a positive $\mathbb{A}$ which can be achieved by choosing the one-electron space in a suitable way, namely by requiring that P_+ is the projector onto the positive spectral subspace of the operator A. It immediately follows that

$$\mathbb{A}=\int d^3x\;:\Psi^*(x)D_0\Psi(x):+\frac{\alpha}{2}\int d^3x\int d^3y\;\frac{(\operatorname{sgn}A)(x,y)}{|\vec{x}-\vec{y}|}:\Psi^*(x)\Psi(y):$$

Now we restrict our search for an operator. On the one hand this is motivated by physics to obtain an operator which looks like a Dirac operator, on the other hand it is useful in order to simplify the situation: We want to have a translationally invariant operator, namely a Fourier multiplier of the form (with $g_{0,1}$ as real functions)

$$A(\vec{p})=\vec{\alpha}\cdot\vec{e}_{\vec{p}}\,g_1(|\vec{p}|)+\beta g_0(|\vec{p}|) \tag{13}$$

Since $\beta^2=\mathbb{1}_4$ and $\{\alpha_i,\alpha_j\}=2\delta_{ij}\mathbb{1}_4$ the signum of A $(\operatorname{sgn}A)(\vec{p})$ is simply given by

$$(\operatorname{sgn}A)(\vec{p})=\frac{\vec{\alpha}\cdot\vec{e}_{\vec{p}}\,g_1(|\vec{p}|)+\beta g_0(|\vec{p}|)}{(g_1(|\vec{p}|)^2+g_0(|\vec{p}|)^2)^{1/2}}$$

Now we are able to determine the equation for the unknown A, which will turn out to be nonlinear.

For the derivation we make use of Fubini's theorem and basic properties of the Fourier transform, especially that the Fourier transform of a product is given by the convolution of the Fourier transforms of the factors.

$$
\begin{aligned}
\mathbb{A} &= \int d^3p\, A(\vec{p}) : \hat{\Psi}^*(p)\hat{\Psi}(p) : \\
&= \int d^3p\, D_0(\vec{p}) : \hat{\Psi}^*(p)\hat{\Psi}(p) : \\
&+ \frac{\alpha}{2}\int d^3x \int d^3y\, \frac{1}{(2\pi)^3}\int d^3p\, e^{-i\vec{p}\vec{x}} \int d^3q\, e^{+i\vec{q}\vec{y}}\, \frac{(\mathrm{sgn}\, A)(x-y)}{|\vec{x}-\vec{y}|} : \hat{\Psi}^*(p)\hat{\Psi}(q) : \\
&= \int d^3p\, D_0(\vec{p}) : \hat{\Psi}^*(p)\hat{\Psi}(p) : \\
&+ \frac{\alpha}{2(2\pi)^3}\int d^3p \int d^3q \int d^3 \underbrace{x}_{\to x-y} \int d^3y\, e^{-i\vec{p}(\vec{x}+\vec{y})} e^{+i\vec{q}\vec{y}}\, \frac{(\mathrm{sgn}\, A)(x)}{|\vec{x}|} : \hat{\Psi}^*(p)\hat{\Psi}(q) : \\
&= \int d^3p\, D_0(\vec{p}) : \hat{\Psi}^*(p)\hat{\Psi}(p) : \\
&+ \frac{\alpha}{2(2\pi)^3}\int d^3p \int d^3q\, \underbrace{\int d^3x\, e^{-i\vec{p}\vec{x}}\, \frac{(\mathrm{sgn}\, A)(x)}{|\vec{x}|}}_{(2\pi)^{\frac{3}{2}}\mathscr{F}\left(\frac{(\mathrm{sgn}\, A)(\cdot)}{|\cdot|}\right)(p)} \underbrace{\int d^3y\, e^{i(\vec{q}-\vec{p})\vec{y}}}_{(2\pi)^3\delta(p-q)} : \hat{\Psi}^*(p)\hat{\Psi}(q) : \\
&= \int d^3p\, D_0(\vec{p}) : \hat{\Psi}^*(p)\hat{\Psi}(p) : + \frac{\alpha}{2}\int d^3p\, (2\pi)^{\frac{3}{2}}\mathscr{F}\left(\frac{(\mathrm{sgn}\, A)(\cdot)}{|\cdot|}\right)(p) : \hat{\Psi}^*(p)\hat{\Psi}(p) : \\
&= \int d^3p\, D_0(\vec{p}) : \hat{\Psi}^*(p)\hat{\Psi}(p) : + \frac{\alpha}{2}\int d^3p\, \mathscr{F}(\mathrm{sgn}\, A)(\cdot) * \underbrace{\mathscr{F}\left(\frac{1}{|\cdot|}\right)}_{\frac{2}{(2\pi)^2|\cdot|^2}}(p) : \hat{\Psi}^*(p)\hat{\Psi}(p) : \\
&= \int d^3p \left[D_0(\vec{p}) + \frac{\alpha}{4\pi^2}\int d^3q\, \frac{(\mathrm{sgn}\, A)(\vec{q})}{|\vec{p}-\vec{q}|^2}\right] : \hat{\Psi}^*(p)\hat{\Psi}(p) :
\end{aligned}
$$

Comparing the first and the last line yields

$$A(\vec{p}) = D_0(\vec{p}) + \frac{\alpha}{4\pi^2}\int d^3q\, \frac{(\mathrm{sgn}\, A)(\vec{q})}{|\vec{p}-\vec{q}|^2} \tag{14}$$

Plugging in the ansatz (13) gives equations the functions $g_{0,1}$ have to fulfill

$$g_0(|\vec{p}|) = m_0 + \frac{\alpha}{4\pi^2}\int d^3q\, \frac{1}{|\vec{p}-\vec{q}|^2}\frac{g_0(|\vec{q}|)}{(g_1(|\vec{q}|)^2 + g_0(|\vec{q}|)^2)^{1/2}} \tag{15}$$

$$g_1(|\vec{p}|) = |\vec{p}| + \frac{\alpha}{4\pi^2}\int d^3q\, \frac{1}{|\vec{p}-\vec{q}|^2}\frac{\vec{e}_{\vec{p}}\cdot\vec{e}_{\vec{q}}\, g_1(|\vec{q}|)}{(g_1(|\vec{q}|)^2 + g_0(|\vec{q}|)^2)^{1/2}} \tag{16}$$

The next chapter is devoted to determining the so called dressed electron and will turn to the question under which conditions this system of coupled equations is solvable. The existence of solutions will be studied by a fixed point argument.

4.2 Dressed electron

In order to find the solution of the system (15)-(16) we will consider a fixed point method. The functions $g_{0,1}$ only depend on the absolute value of the momentum, therefore we can integrate out the angle $\theta = \angle(\vec{p}, \vec{q})$ between $\vec{p}$ and $\vec{q}$. The variables u and v denote the absolute values of $\vec{p}$ and $\vec{q}$ respectively.

$$\begin{aligned}
g_0(u) &= m_0 + \frac{\alpha}{4\pi^2} \int d^3q \, \frac{1}{|\vec{p}-\vec{q}|^2} \frac{g_0(|\vec{q}|)}{(g_1(|\vec{q}|)^2 + g_0(|\vec{q}|)^2)^{1/2}} \\
&= m_0 + \frac{\alpha}{4\pi^2} 2\pi \int dv \, v^2 \int_{-1}^{1} \frac{d(\overbrace{\cos(\theta)}^{x})}{u^2+v^2-2uv\cos(\theta)} \frac{g_0(v)}{(g_1(v)^2+g_0(v)^2)^{1/2}} \\
&= m_0 + \frac{\alpha}{4\pi^2} 2\pi \int dv \, \frac{-v}{2u} \left[\log\left(u^2+v^2-2uvx\right)\right]_{-1}^{1} \frac{g_0(v)}{(g_1(v)^2+g_0(v)^2)^{1/2}} \\
&= m_0 + \frac{\alpha}{2\pi} \int dv \, \frac{v}{2u} \log\left(\frac{(u+v)^2}{(u-v)^2}\right) \frac{g_0(v)}{(g_1(v)^2+g_0(v)^2)^{1/2}} \\
&= m_0 + \frac{\alpha}{2\pi} \int dv \, \frac{v}{u} \, Q_0\left(\frac{1}{2}\left(\frac{u}{v}+\frac{v}{u}\right)\right) \frac{g_0(v)}{(g_1(v)^2+g_0(v)^2)^{1/2}}
\end{aligned} \tag{17}$$

with a function $Q_0(z)$ which for $z > 1$ is defined by

$$Q_0(z) = \frac{1}{2} \log \frac{z+1}{z-1}$$

$$\begin{aligned}
g_1(u) &= u + \frac{\alpha}{4\pi^2} 2\pi \int dv \, v^2 \int_{-1}^{1} d(\overbrace{\cos(\theta)}^{x}) \frac{\cos(\theta)}{u^2+v^2-2uv\cos(\theta)} \frac{g_1(v)}{(g_1(v)^2+g_0(v)^2)^{1/2}} \\
&= u + \frac{\alpha}{4\pi^2} 2\pi \int dv \left[\frac{xv}{-2u} - \frac{u^2+v^2}{4u^2} \log\left(u^2+v^2-2uvx\right)\right]_{-1}^{1} \frac{g_1(v)}{(g_1(v)^2+g_0(v)^2)^{1/2}} \\
&= u + \frac{\alpha}{2\pi} \int dv \left(\frac{u^2+v^2}{4u^2} \log\left(\frac{(u+v)^2}{(u-v)^2}\right) - \frac{v}{u}\right) \frac{g_1(v)}{(g_1(v)^2+g_0(v)^2)^{1/2}} \\
&= u + \frac{\alpha}{2\pi} \int dv \, \frac{v}{u} Q_1\left(\frac{1}{2}\left(\frac{u}{v}+\frac{v}{u}\right)\right) \frac{g_1(v)}{(g_1(v)^2+g_0(v)^2)^{1/2}}
\end{aligned} \tag{18}$$

with a function $Q_1(z)$ which for $z > 1$ is defined by

$$Q_1(z) = \frac{z}{2} \log \frac{z+1}{z-1} - 1$$

The restriction $z > 1$ is no real limitation since the minimum of $\frac{1}{2}\left(\frac{u}{v}+\frac{v}{u}\right)$ is 1 and only attained when $v = u$ which is a set of zero measure.

Now we will solve the system for the case of zero bare mass m_0=0, which is almost trivial and then for the case of non-zero bare masses $m_0 > 0$, which requires a lot more work.

4.2.1 Zero bare mass

If the bare mass m_0 is zero, then g_0 also can be set to zero, which is compatible with equation (17). Plugging this result into (18) one can immediately obtain g_1 by simple integration.

$$g_0(u) = 0 \tag{19}$$

$$\begin{aligned}
g_1(u) &= u + \frac{\alpha}{2\pi}\int_0^\Lambda dv\ \frac{v}{u}\underbrace{Q_1\left(\frac{1}{2}\left(\frac{u}{v}+\frac{v}{u}\right)\right)}_{Q_1(z)=\frac{z}{2}\log\frac{z+1}{z-1}-1} \\
&\overset{\tilde{v}=\frac{v}{u}}{=} u + \frac{\alpha}{2\pi}u\int_0^{\frac{\Lambda}{u}} d\tilde{v}\ \frac{1+\tilde{v}^2}{2}\log\left|\frac{1+\tilde{v}}{1-\tilde{v}}\right| - \tilde{v} \\
&= u + \frac{\alpha}{2\pi}u\int_0^{\frac{\Lambda}{u}} d\tilde{v}\ \frac{1+\tilde{v}^2}{2}\left(\log(1+\tilde{v}) - \log|1-\tilde{v}|\right) - \tilde{v} \\
&= u + \frac{\alpha}{2\pi}\frac{u}{2}\int_0^{\frac{\Lambda}{u}} d\tilde{v}\ (1+\tilde{v})^2\log(1+\tilde{v}) - (1-\tilde{v})^2\log|1-\tilde{v}| - 2\tilde{v}\log|1-\tilde{v}^2| - 2\tilde{v} \\
&= u + \frac{\alpha}{2\pi}\frac{u}{2}\int_0^{\frac{\Lambda}{u}} d\tilde{v}\ (1+\tilde{v})^2\log(1+\tilde{v}) - (1-\tilde{v})^2\log|1-\tilde{v}| \\
&\quad - \frac{\alpha}{2\pi}\frac{u}{2}\int_0^{\left(\frac{\Lambda}{u}\right)^2} \underbrace{d\left(\tilde{v}^2\right)}_{x}\log|1-\tilde{v}^2| - \frac{\alpha}{2\pi}\frac{u}{2}\left[\tilde{v}^2\right]_0^{\frac{\Lambda}{u}} \\
&= u + \frac{\alpha}{2\pi}\frac{u}{2}\left[\underbrace{(1+\tilde{v})^3}_{1+3\tilde{v}+3\tilde{v}^2+\tilde{v}^3}\left(\frac{\log(1+\tilde{v})}{3} - \frac{1}{9}\right) + \underbrace{(1-\tilde{v})^3}_{1-3\tilde{v}+3\tilde{v}^2-\tilde{v}^3}\left(\frac{\log|1-\tilde{v}|}{3} - \frac{1}{9}\right)\right]_0^{\frac{\Lambda}{u}} \\
&\quad + \frac{\alpha}{2\pi}\frac{u}{2}\left[\left(1-\tilde{v}^2\right)\left(\log|1-\tilde{v}^2| - 1\right) - \tilde{v}^2\right]_0^{\frac{\Lambda}{u}} \\
&= u + \frac{\alpha}{2\pi}\frac{u}{2}\left[\frac{4}{3}\log\left|1-\left(\frac{\Lambda}{u}\right)^2\right| + \frac{\Lambda}{u}\log\left|\frac{1+\frac{\Lambda}{u}}{1-\frac{\Lambda}{u}}\right| + \frac{1}{3}\left(\frac{\Lambda}{u}\right)^3\log\left|\frac{1+\frac{\Lambda}{u}}{1-\frac{\Lambda}{u}}\right| - \frac{2}{3}\left(\frac{\Lambda}{u}\right)^2\right] \\
&= u + \frac{\alpha}{2\pi}u\left[\frac{2}{3}\log\left|\left(\frac{\Lambda}{u}\right)^2 - 1\right| + \frac{1}{6}\frac{\Lambda}{u}\log\left|\frac{\frac{\Lambda}{u}+1}{\frac{\Lambda}{u}-1}\right|\left(3+\left(\frac{\Lambda}{u}\right)^2\right) - \frac{1}{3}\left(\frac{\Lambda}{u}\right)^2\right]
\end{aligned} \tag{20}$$

Since we would like that A looks like a renormalized Dirac operator, we are interested in the asymptotic behaviour for small ratios $\frac{u}{\Lambda}$ which is obtained by rewriting the logarithmic term and expanding the resulting logarithms of the type $\log\left|1 \pm \frac{u}{\Lambda}\right| \approx 1 \pm \frac{u}{\Lambda} + \frac{1}{2}\left(\frac{u}{\Lambda}\right)^2$ upto quadratic order.

$$\begin{aligned}
g_1(u) &= u + \frac{\alpha}{2\pi}u\left[\frac{2}{3}\log\left|\left(\frac{\Lambda}{u}\right)^2 - 1\right| + \frac{1}{6}\frac{\Lambda}{u}\underbrace{\overbrace{\log\left|\frac{1+\frac{u}{\Lambda}}{1-\frac{u}{\Lambda}}\right|}^{\log\left|1+\frac{u}{\Lambda}\right|-\log\left|1-\frac{u}{\Lambda}\right|}}_{\approx 2\frac{u}{\Lambda}}\left(3+\left(\frac{\Lambda}{u}\right)^2\right) - \frac{1}{3}\left(\frac{\Lambda}{u}\right)^2\right] \\
&\approx u + \frac{\alpha}{2\pi}u\left[\frac{2}{3}\log\left|\left(\frac{\Lambda}{u}\right)^2\right| + \frac{1}{3}\left(3+\left(\frac{\Lambda}{u}\right)^2\right) - \frac{1}{3}\left(\frac{\Lambda}{u}\right)^2\right]
\end{aligned}$$

$$\approx u\left(1+\frac{\alpha}{2\pi}\left[\frac{4}{3}\log\left|\frac{\Lambda}{u}\right|+1\right]\right)$$
$$\overset{\frac{\Lambda}{u}\gg 1}{\longrightarrow} u\left(1+\frac{2\alpha}{3\pi}\log\left|\frac{\Lambda}{u}\right|\right) \tag{21}$$

Because of this logarithmic divergence for $\Lambda \to \infty$ it is impossible to identify the operator A with a renormalized Dirac operator like in (6).
At this stage we note that the logarithmic term for zero bare mass m_0 looks perfectly similar to the usual charge renormalization in QED, when the bare mass m_0 is replaced by the small momentum u.

$$\alpha_{ph}(\alpha, m_0, \Lambda) = \frac{\alpha}{1+\frac{2\alpha}{3\pi}\log\left(\frac{\Lambda}{m_0}\right)}$$

4.2.2 Positive bare mass

In order to solve the system for positive bare masses $m_0 > 0$ we use an argument based on the Banach fixed point theorem.

Banach fixed point theorem

> Let (X, d) be a non-empty complete ***metric space***.
> Let $T : X \to X$ be a ***contraction*** mapping on X
>
> $$\exists q < 1 : d(T(x), T(y)) \leq q \cdot d(x, y) \; \forall x, y \in X$$
>
> Then the map T admits one unique fixed point x^* in X: $T(x^*) = x^*$.

Therefore it has to be shown that T for $\vec{g} = (g_0, g_1)$ defined by

$$T_0(\vec{g})(u) = m_0 + \frac{\alpha}{2\pi}\int_0^\Lambda dv\, \frac{v}{u}\, \underbrace{Q_0\left(\frac{1}{2}\left(\frac{u}{v}+\frac{v}{u}\right)\right)}_{Q_0(z)=\frac{1}{2}\log\frac{z+1}{z-1}} \frac{g_0(v)}{(g_1(v)^2+g_0(v)^2)^{1/2}} \tag{22}$$

$$T_1(\vec{g})(u) = u + \frac{\alpha}{2\pi}\int_0^\Lambda dv\, \frac{v}{u}\, \underbrace{Q_1\left(\frac{1}{2}\left(\frac{u}{v}+\frac{v}{u}\right)\right)}_{Q_1(z)=\frac{z}{2}\log\frac{z+1}{z-1}-1} \frac{g_1(v)}{(g_1(v)^2+g_0(v)^2)^{1/2}} \tag{23}$$

is a contraction on a suitable metric space, namely a neigbourhood of $\vec{g} = (m_0, u)$ describing the free Dirac operator together with the metric generated by the sup norm $\|\cdot\|_\infty$ on $S_{\delta,\varepsilon}$ for $\delta, \varepsilon > 0$

$$S_{\delta,\varepsilon} := \{\vec{g} = (g_0, g_1) |\; \forall u \in [0, \Lambda] : \vec{g}(u) \in [m_0, (1+\delta)m_0] \times [u, (1+\varepsilon)u]\} \tag{24}$$

The metric space $(S_{\delta,\varepsilon}, \|\cdot\|_\infty)$ defined above is complete since it is a closed subset of the ambient complete metric space $(\mathbb{R}^2, \|\cdot\|_\infty)$ because it consists of a tensor product of closed intervals.
Obviously there can be an interdependence between ε, δ and Λ in order to restrict T such that $T : S_{\delta,\varepsilon} \to S_{\delta,\varepsilon}$. In order to determine this interdependence we have to find upper bounds for g_0 and g_1.

Upper bound for g_0 Since due to (22) obviously $T_0(g)(u) \geq m_0$ as well as $T_1(g)(u) \geq u$ holds one can estimate g_0 in $S_{\delta,\varepsilon}$ from above by

$$\begin{aligned}
g_0(|\vec{p}|) &= m_0 + \frac{\alpha}{4\pi^2}\int_{|\vec{q}|<\Lambda} d^3q\; \frac{1}{|\vec{p}-\vec{q}|^2}\frac{g_0(|\vec{q}|)}{(g_1(|\vec{q}|)^2+g_0(|\vec{q}|)^2)^{1/2}} \\
&\leq m_0 + \frac{\alpha}{4\pi^2}\int_{|\vec{q}|<\Lambda} d^3q\; \frac{1}{|\vec{p}-\vec{q}|^2}\frac{(1+\delta)m_0}{\sqrt{|\vec{q}|^2+m_0^2}} \\
&= m_0 + \frac{\alpha}{4\pi^2}(1+\delta)\int d^3q\; \frac{\mathbb{1}_{\{|\vec{q}|<\Lambda\}}}{|\vec{p}-\vec{q}|^2}\frac{m_0}{\sqrt{|\vec{q}|^2+m_0^2}} \\
&= m_0 + \frac{\alpha}{4\pi^2}(1+\delta)m_0\left(\frac{1}{|\cdot|^2} * \frac{\mathbb{1}_{\{|\cdot|<\Lambda\}}}{\sqrt{|\cdot|^2+m_0^2}}\right)(\vec{p})
\end{aligned}$$

The last expression emphasizes that the integral is a convolution of two functions which are both symmetrically decreasing which implies that their convolution is also symmetrically decreasing (see Lieb [7]). Therefore we can bound this term from above by setting $\vec{p}=0$.

$$\begin{aligned}
g_0(|\vec{p}|) &\leq m_0 + \frac{\alpha}{4\pi^2}(1+\delta)\int_{|\vec{q}|<\Lambda} d^3q\; \frac{1}{|\vec{q}|^2}\frac{m_0}{\sqrt{|\vec{q}|^2+m_0^2}} \\
&= m_0 + \frac{\alpha}{4\pi^2}(1+\delta)m_0 4\pi \int_0^\Lambda dv\; \frac{1}{\sqrt{v^2+m_0^2}} \\
&= m_0\left[1+\frac{\alpha}{\pi}(1+\delta)\operatorname{Arsinh}\left(\frac{\Lambda}{m_0}\right)\right] \qquad (25)
\end{aligned}$$

$$g_0(|\vec{p}|) \overset{!}{\leq} m_0\,[1+\delta] \qquad (26)$$

Upper bound for g_1 In order to obtain an upper bound for g_1 we split the integration region of (23) into $A := \{\vec{q} : |\vec{q}| < 2|\vec{p}|, |\vec{q}| < \Lambda\}$ and $B := \{\vec{q} : |\vec{q}| \geq 2|\vec{p}|, |\vec{q}| < \Lambda\}$.

$$\begin{aligned}
g_1(|\vec{p}|) &= |\vec{p}| + \frac{\alpha}{4\pi^2}\int_{|\vec{q}|<\Lambda} d^3q\; \frac{1}{|\vec{p}-\vec{q}|^2}\frac{\vec{e}_{\vec{p}}\cdot\vec{e}_{\vec{q}}\, g_1(|\vec{q}|)}{(g_1(|\vec{q}|)^2+g_0(|\vec{q}|)^2)^{1/2}} \\
&= |\vec{p}| + \frac{\alpha}{4\pi^2}\int_A d^3q\; \frac{1}{|\vec{p}-\vec{q}|^2}\frac{\overbrace{\vec{e}_{\vec{p}}\cdot\vec{e}_{\vec{q}}}^{\leq 1}\, g_1(|\vec{q}|)}{(g_1(|\vec{q}|)^2+g_0(|\vec{q}|)^2)^{1/2}} \\
&\quad + \frac{\alpha}{4\pi^2}\int_B d^3q\; \frac{1}{|\vec{p}-\vec{q}|^2}\frac{\vec{e}_{\vec{p}}\cdot\vec{e}_{\vec{q}}\, g_1(|\vec{q}|)}{(g_1(|\vec{q}|)^2+g_0(|\vec{q}|)^2)^{1/2}} \\
&\leq |\vec{p}| + \frac{\alpha}{4\pi^2}\int_A d^3q\; \frac{1}{|\vec{p}-\vec{q}|^2}\frac{(1+\varepsilon)2|\vec{p}|}{(g_1(|\vec{q}|)^2+g_0(|\vec{q}|)^2)^{1/2}} \\
&\quad + \frac{\alpha}{8\pi^2}\int_B d^3q\; \underbrace{\left(\frac{1}{|\vec{p}-\vec{q}|^2}-\frac{1}{|\vec{p}+\vec{q}|^2}\right)}_{=\frac{4|\vec{p}||\vec{q}|cos(\theta)}{(|\vec{p}|^2-|\vec{q}|^2)^2+4|\vec{p}|^2|\vec{q}|^2 sin^2(\theta)}\leq\frac{4|\vec{p}||\vec{q}|}{(|\vec{p}|^2-|\vec{q}|^2)^2}} \frac{g_1(|\vec{q}|)}{(g_1(|\vec{q}|)^2+g_0(|\vec{q}|)^2)^{1/2}} \\
&\leq |\vec{p}| + \frac{\alpha}{4\pi^2}(1+\varepsilon)2|\vec{p}|\int_{|\vec{q}|<\Lambda} d^3q\; \frac{1}{|\vec{p}-\vec{q}|^2}\frac{1}{\sqrt{|\vec{q}|^2+m_0^2}}
\end{aligned}$$

$$+\frac{\alpha}{8\pi^2}\int_B d^3q\,\underbrace{\frac{4|\vec{p}||\vec{q}|}{(|\vec{p}|^2-|\vec{q}|^2)^2}}_{\overset{*}{\geq}\frac{9}{16}|\vec{q}|^4}\frac{(1+\varepsilon)|\vec{q}|}{\sqrt{|\vec{q}|^2+m_0^2}}$$

$$\leq |\vec{p}|+2|\vec{p}|\frac{\alpha}{\pi}(1+\varepsilon)\text{Arsinh}\left(\frac{\Lambda}{m_0}\right)+2|\vec{p}|\frac{\alpha}{\pi}(1+\varepsilon)\frac{16}{9}\text{Arsinh}\left(\frac{\Lambda}{m_0}\right)$$

$$= |\vec{p}|\left[1+\frac{50}{9}\frac{\alpha}{\pi}(1+\varepsilon)\text{Arsinh}\left(\frac{\Lambda}{m_0}\right)\right] \quad (27)$$

$$\overset{!}{\leq} |\vec{p}|\,[1+\varepsilon] \quad (28)$$

The $*$-estimate follows from (for $\epsilon > 1$)

$$\left(\sqrt{\epsilon}a-\frac{b}{\sqrt{\epsilon}}\right)^2 \geq 0 \quad\Rightarrow\quad 2ab\leq \epsilon a^2+\frac{b^2}{\epsilon}$$

Applying this on $(|\vec{p}|^2-|\vec{q}|^2)^2$ for $|\vec{q}|\geq 2|\vec{p}|$

$$\begin{aligned}(|\vec{p}|^2-|\vec{q}|^2)^2 &= |\vec{p}|^4+|\vec{q}|^4-2|\vec{p}|^2|\vec{q}|^2\\ &\geq \underbrace{(1-\epsilon)}_{<0}|\vec{p}|^4+\left(1-\frac{1}{\epsilon}\right)|\vec{q}|^4\\ &\geq \left(\frac{1-\epsilon}{16}+1-\frac{1}{\epsilon}\right)|\vec{q}|^4\end{aligned}$$

Optimizing this lower bound with respect to ϵ leads to the desired lower bound for $\epsilon = +4$

$$\frac{d}{d\epsilon}\left(\frac{1-\epsilon}{16}+1-\frac{1}{\epsilon}\right) = -\frac{1}{16}+\frac{1}{\epsilon^2}\overset{!}{=}0 \quad\Rightarrow\quad \epsilon=4$$

$$(|\vec{p}|^2-|\vec{q}|^2)^2 \geq \frac{9}{16}|\vec{q}|^4 \quad (29)$$

Conclusion that $S_{\delta,\varepsilon}$ maps to itself With the requierement that $g_0(|\vec{p}|)$ and $g_1(|\vec{p}|)$ should be in $S_{\delta,\varepsilon}$ one can conclude the following relations

$$(1+\delta)\overbrace{\frac{\alpha}{\pi}\text{Arsinh}\left(\frac{\Lambda}{m_0}\right)}^{:=Y}\overset{!}{\leq}\delta\Rightarrow\delta\geq\frac{Y}{1-Y} \quad (30)$$

$$\frac{50}{9}(1+\varepsilon)\frac{\alpha}{\pi}\text{Arsinh}\left(\frac{\Lambda}{m_0}\right)\overset{!}{\leq}\varepsilon\Rightarrow\varepsilon\geq\frac{50Y}{9-50Y} \quad (31)$$

$$\frac{Y}{1-Y}\overset{!}{>}0\Rightarrow Y\in(0,1)$$

$$\frac{50Y}{9-50Y}\overset{!}{>}0\Rightarrow Y\in\left(0,\frac{9}{50}\right) \quad (32)$$

This proves the following Lemma

Lemma 1

If $Y := \frac{\alpha}{\pi}\text{Arsinh}\left(\frac{\Lambda}{m_0}\right)$ satisfies $Y < \frac{9}{50}$ and if $\delta \geq \frac{Y}{1-Y}$ holds as well as $\varepsilon \geq \frac{50Y}{9-50Y}$, then T maps $S_{\delta,\varepsilon}$ to itself.

Contraction property It remains to be shown that T is a contraction on $S_{\delta,\varepsilon}$.
The following statement is a version of the mean value theorem for differentiable functions in 2 variables in combination with Schwarz inequality

$$|f(x,y) - f(\tilde{x},\tilde{y})| \leq \|\nabla f(\xi,\eta)\|_2 \cdot \|(x,y) - (\tilde{x},\tilde{y})\|_2 \tag{33}$$

with $(\xi,\eta) = (1-\lambda)(x,y) + \lambda(\tilde{x},\tilde{y})$ $\lambda \in [0,1]$.
Applying this to $f(x,y) = \frac{x}{\sqrt{x^2+y^2}}$ one easily deduces

$$\|\nabla f(x,y)\|_2 = \left\|\left(\frac{y^2}{(x^2+y^2)^{3/2}}, \frac{-xy}{(x^2+y^2)^{3/2}}\right)\right\|_2 = \frac{|y|}{x^2+y^2}$$

and therefore

$$\left|\frac{x}{\sqrt{x^2+y^2}} - \frac{\tilde{x}}{\sqrt{\tilde{x}^2+\tilde{y}^2}}\right| \leq \frac{|\eta|}{\xi^2+\eta^2}\sqrt{(x-\tilde{x})^2+(y-\tilde{y})^2} \quad \forall x,y,\tilde{x},\tilde{y} > 0 \tag{34}$$

With this property one can determine the distance between $T(\vec{g})$ and $T(\vec{\tilde{g}})$

$$\begin{aligned}
&\sqrt{[T_0(g) - T_0(\tilde{g})]^2(u) + [T_1(g) - T_1(\tilde{g})]^2(u)} \\
&\leq |[T_0(g) - T_0(\tilde{g})](u)| + |[T_1(g) - T_1(\tilde{g})](u)| \\
&\leq \frac{\alpha}{2\pi}\int_0^\Lambda dv\frac{v}{u}\left[Q_0\left(\frac{1}{2}\left(\frac{u}{v}+\frac{v}{u}\right)\right)\left|\frac{g_0(v)}{\sqrt{g_0(v)^2+g_1(v)^2}} - \frac{\tilde{g}_0(v)}{\sqrt{\tilde{g}_0(v)^2+\tilde{g}_1(v)^2}}\right|\right. \\
&\qquad\qquad \left. + Q_1\left(\frac{1}{2}\left(\frac{u}{v}+\frac{v}{u}\right)\right)\left|\frac{g_1(v)}{\sqrt{g_0(v)^2+g_1(v)^2}} - \frac{\tilde{g}_1(v)}{\sqrt{\tilde{g}_0(v)^2+\tilde{g}_1(v)^2}}\right|\right] \\
&\leq \frac{\alpha}{2\pi}\int_0^\Lambda dv\frac{v}{u}\left[Q_0\left(\frac{1}{2}\left(\frac{u}{v}+\frac{v}{u}\right)\right)\frac{(1+\varepsilon)v}{v^2+m_0^2} + Q_1\left(\frac{1}{2}\left(\frac{u}{v}+\frac{v}{u}\right)\right)\frac{(1+\delta)m_0}{v^2+m_0^2}\right] \\
&\qquad \cdot \underbrace{\sqrt{(g_0(v)-\tilde{g}_0(v))^2+(g_1(v)-\tilde{g}_1(v))^2}}_{\leq \sup_v \sqrt{(g_0(v)-\tilde{g}_0(v))^2+(g_1(v)-\tilde{g}_1(v))^2} = \|g-\tilde{g}\|} \\
&\leq \frac{\alpha}{2\pi}\int_0^\Lambda dv\frac{v}{u}\left[Q_0\left(\frac{1}{2}\left(\frac{u}{v}+\frac{v}{u}\right)\right)\frac{(1+\varepsilon)v}{v^2+m_0^2} + Q_1\left(\frac{1}{2}\left(\frac{u}{v}+\frac{v}{u}\right)\right)\frac{(1+\delta)m_0}{v^2+m_0^2}\right]\|g-\tilde{g}\| \\
&\overset{*}{\leq} \frac{\alpha}{2\pi}\int_0^\Lambda dv\frac{v}{u}Q_0\left(\frac{1}{2}\left(\frac{u}{v}+\frac{v}{u}\right)\right)\frac{(1+\varepsilon)v+(1+\delta)m_0}{v^2+m_0^2}\|g-\tilde{g}\|
\end{aligned} \tag{35}$$

$$= \frac{\alpha}{2\pi}\int_0^\Lambda dv \frac{v}{u} Q_0\left(\frac{1}{2}\left(\frac{u}{v}+\frac{v}{u}\right)\right)\left[\frac{1+\varepsilon}{\sqrt{v^2+m_0^2}}\underbrace{\frac{v}{\sqrt{v^2+m_0^2}}}_{\leq 1}+\frac{1+\delta}{\sqrt{v^2+m_0^2}}\underbrace{\frac{m_0}{\sqrt{v^2+m_0^2}}}_{\leq 1}\right]\|g-\tilde{g}\|$$

$$\leq (2+\varepsilon+\delta)\frac{\alpha}{2\pi}\int_0^\Lambda dv \frac{v}{u} Q_0\left(\frac{1}{2}\left(\frac{u}{v}+\frac{v}{u}\right)\right)\frac{1}{\sqrt{v^2+m_0^2}}\|g-\tilde{g}\|$$

$$= (2+\varepsilon+\delta)Y\,\|g-\tilde{g}\| \tag{36}$$

Since the supremum is the least upper bound the following holds

$$f(x)\leq C \quad \Rightarrow \quad \sup_x f(x) \leq C$$

which proves that

$$\|T(g)-T(\tilde{g})\| \leq (2+\varepsilon+\delta)Y\,\|g-\tilde{g}\| \tag{37}$$

The upper bound in $*$ holds since one can compare the integrands pointwise (t:=v/u)

$$Q_0\left(\frac{1}{2}\left(t+t^{-1}\right)\right) \geq Q_1\left(\frac{1}{2}\left(t+t^{-1}\right)\right)$$

$$\log\left|\frac{t+1}{t-1}\right| \geq \frac{1}{2t}(t^2+1)\log\left|\frac{t+1}{t-1}\right| - 1$$

$$\log\left|\frac{t+1}{t-1}\right| \leq \frac{2t}{(t-1)^2}$$

$$\frac{t+1}{t-1} \leq e^{\frac{2t}{(t-1)^2}} = \sum_{n=0}^{\infty}\frac{1}{n!}\left(\frac{2t}{(t-1)^2}\right)^n \geq 1+\frac{2t}{(t-1)^2}$$

Therefore it suffices to show

$$\frac{t+1}{t-1} \leq 1+\frac{2t}{(t-1)^2}$$

$$1+\frac{2}{t-1} \leq 1+\frac{2t}{(t-1)^2}$$

$$1+\frac{2t}{(t-1)^2}-\frac{2}{(t-1)^2} \leq 1+\frac{2t}{(t-1)^2} \qquad \checkmark$$

This proves the following theorem

Theorem 1

If δ, ε and Y satisfy the conditions of Lemma 1 [namely $Y := \frac{\alpha}{\pi}\text{Arsinh}\left(\frac{\Lambda}{m_0}\right) < \frac{9}{50}$ and $\delta \geq \frac{Y}{1-Y}$ an $\varepsilon \geq \frac{50Y}{9-50Y}$] and $(2+\varepsilon+\delta)Y < 1$, then T is a contraction.

When choosing the lowest possible values for δ and ε

$$\delta = \frac{Y}{1-Y} \qquad \varepsilon = \frac{50Y}{9-50Y}$$

the values for Y fulfilling the contraction property for T mapping $S_{\delta,\varepsilon}$ to itself can be obtained by requiring

$$\begin{aligned}
(2+\varepsilon+\delta)Y &< 1 \qquad Y < \frac{9}{50} \\
2Y + \frac{50Y^2}{9-50Y} + \frac{Y^2}{1-Y} &< 1 \qquad | \cdot (9-50Y)(1-Y) \\
(18Y - 118Y^2 + 100Y^3) + (50Y^2 - 50Y^3) + (9Y^2 - 50Y^3) &< 50Y^2 - 59Y + 9 \\
-59Y^2 + 18Y &< 50Y^2 - 59Y + 9 \\
Y^2 - \frac{77}{109}Y + \frac{9}{109} &> 0 \\
Y_{1/2} = \frac{77 \pm \sqrt{77^2 - 4 \cdot 9 \cdot 109}}{218} &\approx \begin{cases} 1/7 < 9/50 \\ 10/17 > 9/50 \end{cases}
\end{aligned}$$

This proves the following Corollary

Corollary 1

If $Y := \frac{\alpha}{\pi}\text{Arsinh}\left(\frac{\Lambda}{m_0}\right)$ satisfies $Y < \frac{1}{7}$ and if $\delta = \frac{Y}{1-Y}$ holds as well as $\varepsilon = \frac{50Y}{9-50Y}$, then T has a unique fixpoint.

Conclusion of this section is that for small enough cutoffs Λ fulfilling roughly $\ln(\Lambda/m_0) < \alpha^{-1}$ the system determining the operator A can be solved. Actually the constraint is not severe, since for the physical value for $\alpha = 1/137$ the cutoff compared to the bare mass can be huge $\frac{\Lambda}{m_0} \leq e^{18}$. In fact this constraint seems to be an artefact of the used method of computation since the same result was obtained in [3] using a completely different method yielding no condition linking α and Λ. At this stage we can also note that for the fixed point method an in some sense perturbative ingredient was used for computational convenience.

The operator A and the Dirac operator have a similar form, but A is not of Dirac form itself. In what sense this different form can be interpreted as renormalization will be discussed. The next chapter will also analyze if the candidates for the renormalization respect the general properties of possible renormalized parameters.

4.3 Properties of the renormalized Hamiltonian

It has been shown that the bare Hamiltonian $\mathbb{H}^{bare}$ is apart from additive constants equivalent to a renormalized Hamiltonian $\mathbb{H}^{ren}_{\alpha}$ of the form

$$\mathbb{H}^{ren}_{\alpha} = \int d^3x \ : \Psi^*(x) A \Psi(x) : + \mathbb{W}^{ren}_{\alpha}$$

where A is an operator of the form

$$A(\vec{p}) := \vec{\alpha} \cdot \vec{e}_{\vec{p}} g_1(|\vec{p}|) + \beta g_0(|\vec{p}|)$$

For positive bare masses and in the limit of small momenta $g_{0,1}$ are positive and in particular g_0 is constant and g_1 proportional to $|\vec{p}|$. Therefore in this limit the operator A takes approximately the form of a renormalized Dirac operator.

$$D_{Z,m} = Z^{-1}(\vec{\alpha} \cdot \vec{p} + m\beta)$$

The relation between A and $D_{Z,m}$ is given by setting

$$Z^{-1} = \lim_{u \to 0^+} \frac{g_1(u)}{u} \qquad m = Z g_0(0) \tag{38}$$

For physical reasons the relations $m > m_0$ and $Z < 1$ should hold, which is intuitively clear since the physical mass includes the self-energy. The relation $Z < 1$ is clear since $g_1(u) > u$ follows directly from (17). It remains to check if $m > m_0$ which is equivalent to $\frac{m}{m_0} Z^{-1} > Z^{-1}$ meaning that

$$\frac{g_0(0)}{m_0} > \lim_{\vec{p} \to 0} \frac{g_1(|\vec{p}|)}{|\vec{p}|} \tag{39}$$

For the following theorem we will prove more than this, namely

$$\frac{g_0(|\vec{p}|)}{m_0} > \frac{g_1(|\vec{p}|)}{|\vec{p}|} \qquad \forall \vec{p} \neq 0$$

In order to prove this we define a subset $\tilde{S}_{\delta,\varepsilon}$ of $S_{\delta,\varepsilon}$ where

$$\frac{g_0(|\vec{p}|)}{m_0} \geq \frac{g_1(|\vec{p}|)}{|\vec{p}|} \qquad \forall \vec{p} \neq 0$$

holds. It will be shown that T leaves $\tilde{S}_{\delta,\varepsilon}$ invariant which means that the inequality above extends to $S_{\delta,\varepsilon}$. This is due to the uniqueness of the solution on $S_{\delta,\varepsilon}$ arising from the fact that the fixed point argument is also valid on $\tilde{S}_{\delta,\varepsilon}$. T leaving $\tilde{S}_{\delta,\varepsilon}$ invariant is equivalent to

$$\int_0^{\Lambda} dv \frac{v}{u} Q_0\left(\frac{1}{2}\left(\frac{u}{v} + \frac{v}{u}\right)\right) \underbrace{\frac{g_0(v)/m_0}{\sqrt{g_1(v)^2 + g_0(v)^2}}}_{\frac{1}{\left(\frac{m_0 g_1(v)}{g_0(v)}\right)^2 + m_0^2} \geq \frac{1}{v^2 + m_0^2}}$$

$$\geq \int_0^\Lambda dv \frac{v}{u} Q_1\left(\frac{1}{2}\left(\frac{u}{v}+\frac{v}{u}\right)\right) \underbrace{\frac{g_1(v)/u}{\sqrt{g_1(v)^2+g_0(v)^2}}}_{\frac{v/u}{\sqrt{v^2+\left(\frac{v g_0(v)}{g_1(v)}\right)^2}} \leq \frac{v/u}{\sqrt{v^2+m_0^2}}}$$

The estimations above were made using the inequality which determines $\tilde{S}_{\delta,\varepsilon}$ and show that it suffices to show

$$\int_0^\Lambda dv \frac{v}{u} Q_0\left(\frac{1}{2}\left(\frac{u}{v}+\frac{v}{u}\right)\right) \frac{1}{v^2+m_0^2} \geq \int_0^\Lambda dv \frac{v}{u} Q_1\left(\frac{1}{2}\left(\frac{u}{v}+\frac{v}{u}\right)\right) \frac{v/u}{\sqrt{v^2+m_0^2}}$$

This can be proved by comparing the integrands pointwise which is equivalent to $(t := v/u \neq 1)$

$$\begin{aligned} Q_0\left(\frac{1}{2}\left(t+t^{-1}\right)\right) &\geq Q_1\left(\frac{1}{2}\left(t+t^{-1}\right)\right) t \\ \log\left|\frac{t+1}{t-1}\right| &\geq \frac{1}{2}(t^2+1)\log\left|\frac{t+1}{t-1}\right| - t \\ \log\left|\frac{t+1}{t-1}\right| &\leq \frac{2t}{t^2-1} \\ \frac{t+1}{t-1} &\leq e^{\frac{2t}{t^2-1}} = \sum_{n=0}^{\infty} \frac{1}{n!}\left(\frac{2t}{t^2-1}\right)^n \geq 1+\frac{2t}{t^2-1}+\frac{2t^2}{(t^2-1)^2} \end{aligned}$$

Therefore it suffices to show

$$\begin{aligned} \frac{t+1}{t-1} &\leq 1+\frac{2t}{t^2-1}+\frac{2t^2}{(t^2-1)^2} \\ 1+\frac{2}{t-1} &\leq 1+\frac{2t}{t^2-1}+\frac{2t^2}{(t^2-1)^2} \\ 1+\frac{2t}{t^2-1}+\frac{2}{t^2-1} &\leq 1+\frac{2t}{t^2-1}+\frac{2t^2}{(t^2-1)^2} \\ 1+\frac{2t}{t^2-1}+\frac{2t^2}{(t^2-1)^2}-\frac{2}{(t^2-1)^2} &\leq 1+\frac{2t}{t^2-1}+\frac{2t^2}{(t^2-1)^2} \qquad \checkmark \end{aligned}$$

which is clear and also ensures that the strict inequality also holds. This proves the concluding theorem

Theorem 2

If $Y < \frac{1}{7}$ the unique solution of the system (22)-(23) has the property

$$\frac{g_0(|\vec{p}|)}{m_0} > \frac{g_1(|\vec{p}|)}{|\vec{p}|} \qquad \forall \vec{p} \neq 0 \tag{40}$$

5 Interpretation

What we have done so far can be recapped as starting with the bare Hamiltonian $\mathbb{H}^{bare}$ with a Coloumb interaction in analogy to the classical electrostatic energy of a field. After normal ordering we got an equivalent Hamiltonian $\mathbb{H}_{\alpha}^{ren} = \mathbb{H}^{bare}$ (up to a cutoff-dependent constant) for the case of momenta which are small compared to the ultraviolet cutoff Λ. This was made possible by using an α-dependent representation of the canonical anticommutation relations, which changes the notion of one-electron states and therefore the field Ψ.
Since the two Hamiltonians are on a formal level the same, the main difference is that in the renormalized Hamiltonian a main part of the electrostatic energy was migrated to the leading term containing an operator of Dirac type.

$$\int d^3x \; :\Psi^*(x) D_{Z,m} \Psi(x):$$

The remaining part $\mathbb{W}_{\alpha}^{ren}$ is now supposed to be less important than the original one. Since the renormalized terms are all normal-ordered they vanish on one-electron-states defined by $\Psi^*|0\rangle$, with the vacuum of the new Ψ.
Neglecting the interaction term $\mathbb{W}_{\alpha}^{ren}$ the Hamiltonian is still not given by a Dirac operator, but a Dirac operator with a prefactor Z^{-1} we called wavefunction renormalization.

$$D_{Z,m} = Z^{-1}(\vec{\alpha} \cdot \vec{p} + m\beta)$$

Here we have a renormalization of an operator giving rise to different possible interpretations how to adress this fact.

Field renormalization $\Psi \to Z^{-1/2}\Psi$ The factor Z^{-1} could be absorbed into the field by taking $\Psi \to Z^{-1/2}\Psi$, but we have to note that this would also change the anticommutation relations. In addition in this interpretation the mass renormalization reads as $g_0(0)/Z$ and the charge renormalization as e/Z.

Mass renormalization $m \to Z^{-1}m$ Another point of view would be to regard $Z^{-1}m$ as the physical mass, which is reasonable since $Z^{-1}m > m_0$, this requires to change $Z^{-1}\vec{p}$ into $\vec{p}$ using a unitary transformation. Since this corresponds to a change of length scale it amounts to a change in the speed of light which is not universal, but depending on the particle. In this case there is in contrast to the last one no need for a charge renormalization.

Charge renormalization $e \to Z^{-1/2}e$ It is also possible to regard the factor Z^{-1} as a multiplier of the Hamiltonian itself. This requires a renormalization of $\alpha \to Z\alpha$ meaning that the charge is renormalized by $e \to Z^{-1/2}e$. Since the Hamiltonian is the generator for time translation this is equivalent to a change of time scale which is again not universal.

References

[1] LIEB, Elliott H.; SIEDENTOP, Heinz: *Renormalization of the regularized relativistic electron-positron-field* Communications in Mathematical Physics, Vol. 213 (2000), 673-683

[2] BACH, Volker; BARBAROUX, Jean-Marie; HELFFER, Bernard; SIEDENTOP, Heinz: *On the stability of the relativistic electron-positron field* Communications in Mathematical Physics, Vol. 201 (1999), 445-460

[3] HAINZL, Christian; LEWIN, Mathieu; SOLOVEJ, Jan Philip: *The Mean-Field Approximation in Quantum Electrodynamics. The no-photon case.* Communications on Pure and Applied Mathematics, Vol. LX, 0546-0596 (2007) Wiley Periodicals, Inc.

[4] HAINZL, Christian; SIEDENTOP, Heinz: *Non-perturbative mass and charge renormalization in relativistic no-photon quantum electrodynamics* Communications in Mathematical Physics, Vol. 243 (2003), 241-260

[5] WEINBERG, Steven: *The Quantum Theory of Fields Volume I*, Cambridge University Press (2005), 191-255

[6] ZEE, Anthony: *Quantum Field Theory In A Nutshell*, Princeton University Press (2003), 145-150, 183-190

[7] LIEB, Elliot H.; LOSS, Michael; RUSKAI, Mary Beth: *Inequalities: selecta of Elliot H. Lieb*, Springer Verlag (2003), p. 353 Lemma 2.2

[8] FEFFERMAN, C. ; DE LA LLAVE, R. : *Relativistic Stability of Matter-I* Revista Matematika Iberoamericana, Vol. 2 Nr. 1/2 (1986), 119-213